"What If" and its Power to Solve Problems

Arpi Der M.

Copyright © 2012 Arpi Der M.

All rights reserved.

ISBN: 1494764873
ISBN-13: 978-1494764876

DEDICATION

"For my entire life people around me did not hesitate to help and guide me to where I am today. Now, I am dedicating my entire life to help others. There is always something that keeps us inspired and gives us energy to continue on. During the past 10 years of my life on this earth, I have learned three things which keep inspiring me continuously. Learning, experiencing and solving problems. This is what I live for."

RP D.

CONTENTS

	Acknowledgments	i
1	Speak True	1
2	Mayreni	7
3	Biology Zone	11
4	Academic Glossary	17
5	AG - API	24
6	AG - APP	35

Special thanks to Professor De Carvalho for editing this book.

ACKNOWLEDGMENTS

Mom, thank you for all your support regarding to any of my crazy ideas. Even though, you could not understand any of those technical programming conversations that we had, you supported me and gave me the ability to learn the ideology behind coding from two of the best programmers in Iran. You made it possible for me to understand what I want to do for the rest of my life. For that I am thankful.

Dad, thank you for all the electrical components that you gave me to explore. Thank you for your passion to teach me your knowledge as an electrical engineer. Because of you, I became a girl how started to figure out the functionalities of electrical components and how they work since elementary school. Also, I am so thankful that you took me to camping every weekend since I was in kindergarten. Because of those camping days in the middle of desert, I am fearless and for any problems that I face, I can overcome them without being panicking out.

My dear brother ... we had our good and bad times. I clearly remember that you would lock your computer desk with key so I could not use the computer and break it. I was only 12 at that time. But when you were not home, I would open the door with dad's tools so I could use the computer. I was curious what was inside that box that you were locking down. For that particular action of yours I am thankful. Your action, forced me to fix whatever I was breaking on your computer before you could notice. It helped me to look for answers on my own and learn anything about computers without relying on someone else.

All my teachers in Kananian School; Dear Mrs. Rima, Mrs. Anahid, Mrs. Karineh , Mrs. Seda , and Mrs. Caroline, you understood me and supported all of my ideas since I came to that school. You gave me the opportunity to get in to Kharazmi Competition and actually get to higher levels. You inspired me in any way you could. I am thankful for any single day that I was in that school. Now, part of my life's purpose is to continue whatever you have started.

Dr. Jenifer Earl, you supported me in anyway you could more than I imagined. I remember the first time that we met in your office. From that day on you did everything you could to support me with my projects. You gave me a huge opportunity to help Mr. Pinsker to solve technological issues in our school. You opened a huge door of opportunity for me. At the time that I was lost in my thoughts and I could not find a way to express my ideas, you were there for me. Throughout these years, in Hoover High School I was able to achieve many things because of your help and inspiration.

Mrs. Chan, you made me to love what I have hated since I got in elementary school. you made me to love Biology. To be honest, you are the first Biology teacher in all those school years who was able to inspire me to learn biology. What you have done, helped me to develop the Biology Zone app. What you have done for me, took me to another level in my life. You helped me to realize that there is nothing in this world to hate. Since I took your class, I never hated anything. Since, I took your class, Biology became an interesting part of my life. I am so thankful for everything that you have done for me.

Mr. Pinsker, you are the best art teacher that I ever had. You helped me to combine my logic and art together in order to come up with something interesting. You helped me to appreciate art as much as I appreciate technology and logic. You helped me to understand that art and technology are completing each other. Because of you, I was able to provide my codes with attractive user interfaces and to make them understandable and functional for anyone. Also, I am thankful for all those days that we were walking around school to fix technological issues. I have learned a lot from what I have experienced with you as a technology coordinator

Mrs. Mariette Gharakhanian, I don't know how to thank you for everything you have done for me. You helped me to find all the answers for my questions. You made me to relax and think clearly when I was under a great stress and pressure. I am thankful for all of our 15 minute conversations that we had during 5th period break. Your calmness, made me calm every time that I was in pressure. All your words inspired me and gave me even more energy to never give up. If it wasn't for your kindness and passion for helping students, I would not survive all those school days. I clearly remember when I was working on a project and I was getting tired of all those challenges, you would convince me to take breaks and get relaxed. I am thankful that you were there for me when I was lost because of not having any answers for my questions.

Professor De Carvalho, you are the first English Professor that I had in college. Because of your class, I got enough passion to write an entire book in English. Because of your help, I was able to finish this book. Also, I want to thank you for all the time you have put in order to read and edit my work. You helped me to overcome this challenge. I am so glad that I took your class, because it changed the way I was thinking about writing.

SPEAK TRUE
DECEMBER 2008

Problem:

It all started from a single issue. I was sitting in my Armenian class in Kananian School, in Isfahan, Iran. While I was talking to my friend, one of the students pronounced an Armenian world incorrectly. At that moment, my Armenian teacher got mad. She was angry because mispronouncing words is a common issue in our community. For that reason only, our teachers were trying to cope with this problem by printing out lists of mispronounced words and their correct pronunciations. Usually teachers would give these list of words to students, or sometimes they would post them on walls. Sadly enough, what they tried did not help, and students were not paying attention at all. This problem had to be solved. This problem started to live inside my brain and distracted me easily. Therefore, I had to find a solution for it.

Idea:

To solve the problem, I asked myself what if we could have a search engine app for mispronounced words. Teachers would be able to add mispronounced words continuously into that engine and students would be able to search for pronunciation of any word that they had doubts about. This idea gradually implanted itself in my mind until I was able to adapt it in a way that it would become a real solution for the problem.

Inspiration:

The only negative side about this idea was that I did not have the slightest clue about programming. How its done, where I should start from, how I should start and so on. It was a hard situation because for the first time I had an idea that would help to solve a problem. I could not leave that idea behind and move on. It was my first real idea for solving a serious problem. So I had to do something about it. It was a cloudy and rainy Monday morning. We had geography class on the second floor of our school. As usual I went to class to search for something new , but had no success. As soon as the teacher came in, she started to talk about nationwide competition called Kharazmi between students in all schools around the country. This sounded like a normal research competition until I had the paper in my hands. Just holding it gave me an impression that I had found something that I had been looking for few years. Therefore, I started to read it. There was only one rule in that competition. We had to invent something that didn't exist. This rule grabbed my attention for the whole week. The next day, instead of going to school to search for something new, I went to schools to research this competition. That day I felt different, I felt abnormal. This feeling turned me upside down in the way that teachers could notice easily. As soon as I told my school principal that I wanted to participate in this competition no matter what, she started to look to other teachers and they looked back. For a moment I didn't know what was going on until one of them said, I am the first student from our school participating in this competition in 10 years. For another week I didn't know what I was going to work on, until I remembered the pronunciation problem. As soon as I remembered that problem, this changed my life immediately. This realization changed everything and defined a whole new path for me.

What is Speak True?

The main purpose and function of Speak True is to help users check for the pronunciation of words that they are not sure about. For example, if they want to check for the correct pronunciation of a word that they know, they can type in the word and see both wrong and correct pronunciation. Also, it provides an auto complete search engine. For example, if the user types a letter, the search engine lists all the words containing that letter in alphabetical order.

Documentation for Non-developers

Tools for Development:

The original app uses Visual Basic programming language in Visual Studio development environment. Visual Basic is one of the .net programming languages. Programmers back then were using this language to develop different software for Microsoft Windows. Visual Studio is just a development environment from Microsoft that helps to develop and test apps in it.

Research:

Before the development of Speak True, I had to look for all the possible words that Armenians mispronounce. I had to ask all the teachers and Armenian language researchers to provide me with a list of all correct and incorrect words. It took me 2 months to gather the necessary information and organize it in an xml document ready for development of the database.

Process:

The process of the app was fairly simple. It was the idea that made it interesting and eye catching. The codes contained only two text fields and an integrated autocomplete function and a search button that contained a matching function to return the words that matched to the user input.

Documentation for Developers

I have to admit back then I was anything but a developer. I did not know a single thing about programming. This was my first idea and I wanted to develop the app at any cost. Thus, I asked a programmer to help me with the coding. The only coding that I did was to develop the database using XML.

Database Development Process:

XML back then was a common way of storing data. Also, it was a simple markup language. The code that I wrote to store and fetch the data in Speak True software was something like the following:

```xml
<words>
    <word>
            <correct>Correct word 1</correct>
            <wrong>Wrong Word 1</wrong>
    </word>
    <word>
            <correct>Correct word 2</correct>
            <wrong>Wrong Word 2</wrong>
    </word>
    <word>
            <correct>Correct word 3</correct>
            <wrong>Wrong Word 3</wrong>
    </word>
</words>
```

Simulation On Laravel:

In order to respect the simplicity of regional Speak True codes and to show the beauty of the idea with that simple code, I tried to simulate the same app on website using Laravel Framework. I used a very basic MYSQL database and simple search engine with table view. Since back then I only used xml to develop the database, I will only describe how I developed the database in this simulation.

Simulated Database:

For the development of this MYSQL database, I used Seeds bundle for Laravel PHP framework. Basically the structure is very similar to XML, which is why I used it. Seeds takes all the inputted data and plants the data inside MySQL database on the server. The following code represents a basic structure of Seeds.

```
public function grow()
{
    $speaktrue = new Trueterm;
    $speaktrue -> wrongTerm = 'wrong term';
    $speaktrue -> trueTerm = 'true term';
    $speaktrue -> save();

    $speaktrue = new Trueterm;
    $speaktrue -> wrongTerm = 'wrong term 2';
    $speaktrue -> trueTerm = 'true term 2';
    $speaktrue -> save();

    $speaktrue = new Trueterm;
    $speaktrue -> wrongTerm = 'wrong term 3';
    $speaktrue -> trueTerm = 'true term 3';
    $speaktrue -> save();

}
```

Wireframes:

The following are sample wireframes of the website coded with Laravel and designed with Twitter Bootstap.

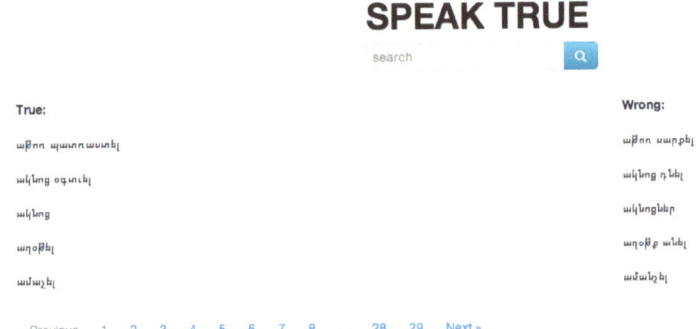

Final Product:

Speak True is available on Mayreni.pagodabox.com website. It is the simulated version of the regional app.

Credits:

I have to thank my Armenian teachers in Iran for this project. Without their help and support I could not gather the necessary resources for this app. Also, I have to thank Mr. Sasoon Amirchian for his huge help with the programming par of this app.

Conclusion:

During this experience, I had to learn how to code. Only two lines of code were needed to change the entire pathway of my life. With learning two lines of code, I decided to continue on and find more problems to solve with the use of technology and codes. This experience proved me that impossible is impossible and everything else is simply possible.

MAYRENI
AUGUST, 2009

Problem:

While trying to accept the fact that I just left the only country that I knew as my home, I started to see an upcoming problem that I knew I would face no matter what. As a matter of fact many people in my situation are facing the same problem. The truth is that we leave our home country without knowing that we are in the danger of forgetting our culture. To be honest while going to a school specific for our culture, I never understood what is it like to forget a culture in our houses. If we don't know the songs, writings, famous people from our culture, if we don't have any art dedicated to our culture, how can we say we are part of that culture. All these problems can be cured. But they have to be cured in a right place and right environment. These problems can be cured starting with learning. But how people outside of our home country can learn about our culture if there is no resources not advanced resources anyway. How kids are supposed to know about their culture if they are living in a foreign country. Schools are considered the best place for curing this problems. However if resources are out dated or they are limited to a very basic level of knowledge, this will not help to solve this problem. Some one has to start a different path. A path that will be effective and will make people to be curious about their culture. Since I was in the danger of losing my culture, I tried to develop a tool based on my knowledge and passion so maybe I can cure this problem starting with my self.

Idea:

Soon, I asked myself "What-If" I could develop an app to learn more about Armenian Culture in a whole new level. I decided to develop this app with c# programming language and WPF which I just got familiar to while exploring. I decided to develop an app while learning about programming. The Idea was to put together a selection of good writers and poets in the app including their biography and their works. But this app instead of showing pure text was going to give the option to listen to both biography and the poems with full inspiring background music. This was going to be a multimedia app that would give the edge to seek for more knowledge about Armenian Culture.

Inspiration:

In this case, I had the inspiration with me just before I was getting ready to leave my home country. This inspiration was a big one and it helped me to do anything in my power to see this project becoming functional. The day before my flight to Vienna, I invited all my teachers and staff members to gather up for a last time. We had our fun and just before saying goodbyes, they gave me a gift. A gift that did not have much of moneywise value but it became the most valuable object that I was going to carry it on while traveling. It was a picture of one of our churches in Isfahan on ceramic. It was the representative of our culture. Perhaps I could say the first one that I had. I never thought a small gift like this would become so inspiring and valuable to me. I was happy that I could carry it on with me. So I secured It like it was a gold. After arriving to Vienna and settling in, the first thing that I did was to hangout that ceramic on its frame to our apartment's wall. It was there for the entire 6 months that we were in Vienna. For each look, I was getting energy. This single gift gave me enough energy and inspiration to not give up what I was going to fight for.

Resources:

The only resources at my disposal at that time, were a notebook full of c# programming language notes, and an Armenian book. During the development of this app, I started to look for recordings of selected poems that I was going to put in the app. However, since I did not find any, I started to ask people around me to help to find people who could do the recordings. Soon enough I found two people who accepted my proposal and did the recordings for both poems and biographies. Perhaps this was the biggest resource that I had which made the app more functional.

Technology Used:

Due to lack of experience in programming field, I was limited to few technologies that my private teachers informed me of. For this app I used C# programming language, Windows Presentation Foundation or (WPF), Microsoft Expression Blend Windows Software, and of course Microsoft Visual Studio development environment.

"What If" and its Power to Solve Problems

Documentation For Non Developers:

This windows application was the second programming experience that I had. Therefore, I was not professional yet. However, compared to the first time I was using more complex codes.

At the very first step I had to tell the application to list of poets for both English and Armenian languages. In order to redirect to the specified page I had to trigger actions for each button click.

In the writers page I had to create sub categories for each writer, and within each writer's page I had to create another set of subcategories for their biography, pictures, and famous works. All these actions are triggered in the same way only with different names so the application will not get confused which function to trigger on user actions.

For the biography pages, I had to initialize a special text view window, which would give users the option to zoom, slide, and change pages. This special window has even the ability to provide search option within the text.

For the poems, I had to use simple views, because I had to provide the option to play, pause, or stop the media player. The media player gives the users the option to listen to the poetry instead of just reading it. In this way it will be more inspiring and clear.

Like I said earlier due to the lack of proficiency and experience, I used the same procedure for the entire app with different names and conditions.

Documentation For Developers:

The codes that I have used for this app, include only button click actions to show the specified pages, and also to control the media player. All the codes are repeated in the entire app with different variable names.

These code blocks are the following:

```
private void Button5_Click(object sender, RoutedEventArgs e)
{
    new zoryan3().Show();
}
```

The above code block is used to show different pages on different button clicks. For example, the one above shows one of the poems of the specified poet.

The following code blocks are used to control the media player. In the following code blocks, each function is responsible for triggering one action. The first one is responsible for Pausing the media. Second function is responsible for Playing the media, and the third one is responsible for stopping the media.

```
private void pause(object sender, RoutedEventArgs e)
{
    this.media.Pause();
}

private void play(object sender, RoutedEventArgs e)
{
    this.media.Play();
}

private void stop(object sender, RoutedEventArgs e)
{
    this.media.Stop();
}
```

As you see there is not much of codes used in this app because most of it is generated automatically when I implemented the components using user interface instead of adding them programmatically. To be honest the tools on user interface and their settings panel helped me to actually build this app.

Final Product:

The final product is accessible from the following link.
https://github.com/arpi6/Mayreni-Windows-App

Further Goals:

Obviously there are further goals for this project. Eventually, I want to expand this application to a set of applications for different schools and different ages with the help of other professionals who have more resources and who are willing to help to teach what is Armenian Culture.

Credits:

I should thank to each of the following people for their support and involvement in this project. Without these people, my work could not become even a prototype of what was on my mind.

Sound Recordings:

Lucineh Boghozian

Maria Asadourian

Additional Support:

Masis Dermardirousian

Caroline Dilanchian

Vache Markarian

BIOLOGY ZONE
JANUARY 2011

Problem:

The problem started to show itself when I took Biology class in High School with English as my third language. I used to go to class and find it challenging to learn and understand all those Biology terms and concepts. For a moment I thought the problem was from me. However, after time was passing and I looked around, I realized that almost 80% of that class was having the same problem. When I started to look around carefully, It became clear that language is our main problem. But somewhere there has to be an answer to this problem. So I started to try harder to get better grades. The most I could get was 90%. Even with getting 90% on tests, the problem was still there. So I had to do something about it. This problem was duelling on my mind for 3 months. The only hope that we had was the hard work of our Biology teacher and of course our teacher assistant. They were doing anything in their power to help us. However, the facts were there. We had difficulty in English and we had to do something about it without losing our chances to learn more.

Idea:

When the winter break came, I asked myself "what-if" there was a self learning application that would help students to learn Biology with using models, graphics, and animations. This question gave me the idea to develop an application that would do exactly the same. I knew that the development of this app would take some time. However, If I wanted to get good grades I had no such time. So I had to move quickly. Since I had not a lot of experience in programming, It would be very challenging and time consuming for me. Even though I had the Idea, I was not sure of my commitment, because I was not trusting myself entirely.

"What If" and its Power to Solve Problems

Inspiration:

I carried on that fear inside me, until I saw the facial expressions of our teacher and teacher assistant while trying so hard to explain us the biology concepts. From seeing their expressions and their tone of voice, anyone would love to study Biology. In that class I have learned that no one is perfect, and not all the experiences come at once. Therefore, if I wanted to solve this problem, I had to take action immediately without any doubts. I realized that if our teachers are trying so hard, then we have to try harder to learn what they are teaching us. I also understood that, it does not matter how we learn. Someone can learn with writing, another might learn with reading, and another might learn with watching videos. No one is forced to learn in traditional way. For me, I found out that I would learn with mixing the given information in class with my passion for coding and programming. This is when I decided to take action without doubting for a single moment. It may seem simple, but the expressions on my teacher's and teacher assistant's faces while trying so hard within their power to help us, gave me all the necessary inspiration to take this step. An inspiration was all I needed to convert this idea to an action and to help students like myself to Learn Biology or any other course without language being a barrier.

I had the Idea, I had the inspiration, and I had a little experience in programming. These were all the tools that I had to convert this idea to a solution. Therefore, I started to expand my knowledge about C# programming language in order to develop the Biology Zone application.

Resources:

For this app the only recourse that I had was my Biology binder. In the binder, I had all the notes, quizzes, diagrams, and pictures that our teacher gave us. That binder had everything I needed to start the development of this app.

Technology Used:

Since I only knew how to program with c# programming language, I had to choose the necessary technology based on that particular programming language. After researching for a while, I decided to use **WPF**, **XAML**, with **C#** and **.NET Framework** to develop a program for Microsoft Windows.

Tools:

I guess in programming everything is related. In order to program for Microsoft windows with the specified languages and framework, I needed a development environment to write the codes in it. Again after a while of research, I found out that **Microsoft Visual Studio** and **Microsoft Expression Blend** development environments are the best options.

Documentation For Non Developers:

Developing The App:

Since I had not a lot of experience in programming, I had to keep the codes simple. In another words, I had to achieve my goal with writing long but simple codes instead of short and complex ones. In order to develop the app, I started to sketch the app like a storyboard on paper. The app itself did not have much of complex codes. I just had to write the functions for navigating through pages and controlling animations. For the rest of the app, I had to type all the texts inside pages and put the pictures on the desired locations. Basically, I built the app with use of 5 methods and functions only.

Documentation For Developers:

In order to develop an app, it is clear for all the developers that a concise plan is needed. In order to develop my app, I started to categorize all the information that I wanted to represent in the app. Therefore, I came up with this diagram which represents the flow of the data. Basically I applied this diagram to each of the chapters that I wanted to include in the app.

Sketches:

After determining the entire map and flow of the app, I needed to determine how to develop each page. From where to start and where to end. Therefore, based on the little habit of developers, I decided to sketch an outline of each page on a paper with a pen. This sketching helped me to figure out where to put each code block and how to link it to the other parts.

Developing The App:

As mentioned earlier, the code blocks for each page are super simple. They only handle the navigation and any click action. However, before beginning to write the codes for each page, I had to make sure that I included all the header files. Since header files refer to different libraries, the absence of one single library would cause the code blocks to malfunction.

The following are the imported libraries that I included in the header of each file.

```
using System;
using System.Collections.Generic;
using System.Text;
using System.Windows;
using System.Windows.Controls;
using System.Windows.Data;
using System.Windows.Documents;
using System.Windows.Input;
using System.Windows.Media;
using System.Windows.Media.Imaging;
using System.Windows.Shapes;
```

After including all the required libraries, I just had to link the button events from each chapter to their sub categories. Based on my diagram, subcategories for each chapter are "Notes", "Test yourself", and "Glossary". Of course based on the material, some chapters might have "Diagrams" sub category. The code for adding button click event to redirect to these pages is the following.

```
private void btnnote_Click(object sender, System.Windows.Routed EventArgs e)
{
    note1 note = new note1();
    note.Show();
}
```

After adding event handlers of sub categories for each chapter, I had to handle the events on each subcategory controller page. Some sub categories had few events and some did not have any. It all depends on the content and information.

As mentioned earlier due to lack of experience in this field, I had to use long yet simple ways to get things done. For instance in order to make the glossary functional I had to search for specific key terms with what users write in the text field The following code block represents a small part of the code that I wrote for this glossary.

```
private void btnsearch_Click(object sender, System.Windows.Rout
edEventArgs e)
{
    if (txtsearch.Text == "genetics")
    {
        txtdefinition.Text = (" Scientific study of heredity. w
hich is how offspring inherits traits from parents.");
    }

    if (txtsearch.Text == "trait")
    {
        txtdefinition.Text = (" A Characteristic that can be pa
ssed from parent to offspring. (Ex: Eye-Color, Hair-
Color, etc..)");
    }


}
```

What this code does, is that it gets the text entered in the text field by users. After getting the text, I used if statements to match the word entered with my individual keywords in this particular glossary. Yes it is right I wrote 270 if statement blocks for the glossary part of this app. If the text in the field matches a certain keyword, the definition text field will show the corresponding definition.

Well as I mentioned earlier this was pretty much all the code that I used over and over again for each chapter and its sub categories. Not much, but more complex than the previous projects.

Final Product:

The BiologyZone windows app is available in a Github repository. This codes are public and anyone can get the codes. This app is a prototype and right now it is not in an installable wizard form. Any one can open the app with having .NET Framework 4.0 in their computer. The link for the repository is, https://github.com/arpi6/BiologyZone_1.

Further Goals:

The main goal for this prototype is to separate all the subcategories and develop them in a series of mobile, and tablet apps. I decided to break this app into smaller apps due to the load of information. This way different groups can focus on updating a single app. Also, another purpose of this breakdown is that a lot of students will get the opportunity to work on this project with gaining real life experiences while helping to change the traditional way of learning into an interesting, and more effective one.

Credits:

For the resources, support, and inspiration of this project I should thank Mrs Cynthia Chan, Biology teacher in High School. Dr Jennifer Earl , the principal of Hoover High School. And finally, Mrs Mariette Gharakhanian for her outstanding inspiration to finish this prototype. Without them, this project would not be as effective and functional as anticipated.

ACADEMIC GLOSSARY
OCTOBER 2012

Problem:

In Fall Semester, 2012, I used to work in the mornings until 12:00 pm and then I had English class at 1:00 pm . Every time that I had a test, I would start turning the pages of my English book until I could find the definition of a specific word. Since, I had only 30 minute to review for the tests, I was finding it extremely challenging to turn all those pages in order to find a single word. This challenge made me eager to look for a reasonable solution.

Idea:

After having the same problem for almost every single test, I thought "What if" I there was a website application for all the glossary words in our academic books. "What if" that website had a search engine to search for words.

Inspiration:

In MINEA USA LLC as a web developer I had to write web applications for different projects. During one of our projects, we started to find problems in developing with pure PHP codes. After a while, I saw a post "Why is LARAVEL taking the PHP community by storm." The next day I went to work and my boss asked me if I watched the videos for cake PHP frameworks." When I answered him about a post I saw for Laravel, he agreed that it was a way better framework than cake PHP. As a result, I started to watch all the video tutorials for LARAVEL. After a while, when we were testing the framework, suddenly we noticed that I had created a whole system for our project. From that day on, I couldn't wait until mornings so I could go to work and have fun while playing with those codes. Every single time that we used to run a block of code for a test, we held our breath until we could see the results. If it was successful, we would scream and laugh like winners. If not we would continue testing until it worked. Working with LARAVEL opened a lot of doors for my web based projects, including The ACADEMIC GLOSSARY project. Therefore, I started to think of actually finishing this project as a web application.

Detailed Process

Tools:

In order to set up this project, a couple of development tools and languages were used. These include Laravel framework for PHP, Twitter Bootstrap, MySql, Github, and Pagoda Box.

Developing the Database:

In order to design and develop the database for the Academic glossary, the MySql database which is a very common database in php based app developments is used. In the database each glossary has its own separate table. Each table consists of a few fields. The fields include id, term, and definition. The database schema itself is created by migrations in laravel. Therefore, on any development machine, developers can migrate the database on command line tool and have access to the database. Also, the data entries are not entered in the usual way. In this project the "plant" bundle is been used in laravel so the words and definitions can be seeded in the database with the **"plant seed::All command"**. In this way, any collaborating developer will have access to the data.

Developing the Application:

The development of the application itself is done with Laravel framework for php language. Laravel makes the core of this application more organized and powerful. In order to start developing the application, first it is critical to set the configuration based on database and server credentials. The first and most important setting is the application key inside application configuration file. Each app should have its own application key, without which the application will not run. Then it is necessary to set the database credentials inside database configuration files. These are necessary steps before the application is developed. Laravel uses its own templating engine called "Blade". Blade is used inside the view files. Also, there is a folder called Controller, which includes all the controller files. Controller files are responsible for holding individual functions related to database models, actions, and pages. Since this application does not have a complex structure, there is only one universal controller file in use. This controller is for handling the "About" and "Glossary" pages. The other controllers belong to each individual object (i.e. Biology Glossary). The models, migrations and seeds files also follow the same concept. Each glossary has its own seed, model, view, and migration files. This is a fairly basic structure for a Laravel framework application.

Designing the Application Using Twitter Bootstrap:

When the core of the application becomes functional, it is necessary to design the interface with using markup languages like HTML and CSS. In this particular application, the Twitter Bootstrap templating framework is used to power up the html design. Bootstrap adds a responsive interface to the application.

Deploying the Codes to the Production Environment:

After designing the interface and powering it up with Bootstrap, it is critical to test the application in the development environment. Until all the issues are fixed, the application is encouraged to stay in development mode. After the testing phase, the application needs to be deployed in the production environment. In this particular application, the Pagoda Box server is used. Unlike the traditional way of uploading the codes to the server using "FTP", the "Pagoda Box" deploys the codes when the changes are pushed to the "Github repository". In this way, developers can switch back and forth between different "git commits". Another advantage of Pagoda Box is that, the application can have few webs with cpu and ram. Also, it can be caffeinated to keep the application alert all the time.

Wireframe:

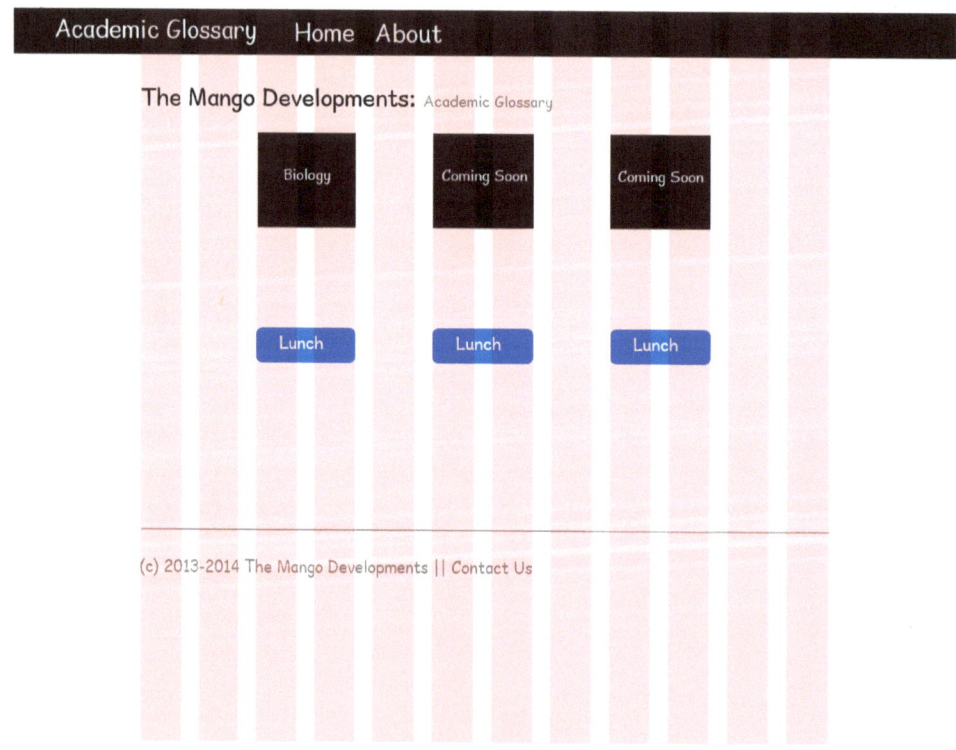

Source Codes:

Database Migrations:

```php
class Biology_Terms_Table {

    /**
     * Make changes to the database.
     *
     * @return void
     */
    public function up()
    {
        Schema::create('biologyterms', function($table) {
            $table->increments('id');
            $table->string('term');
            $table->string('definition');
            $table->timestamps();
        });
    }


    /**
     * Revert the changes to the database.
     *
     * @return void
     */
    public function down()
    {
        Schema::drop('biologyterms');
    }

}
```

Database Collections:

```php
class Biology_Terms_Table {

    /**
     * Make changes to the database.
     *
     * @return void
     */
    public function up()
    {
        Schema::create('biologyterms', function($table) {
            $table->increments('id');
            $table->string('term');
            $table->string('definition');
            $table->timestamps();
        });
    }


    /**
     * Revert the changes to the database.
     *
     * @return void
     */
    public function down()
    {
        Schema::drop('biologyterms');
    }

}
```

Biology Terms Seeds:

```
public function grow()
  {
        $biology = new Biologyterm;
        $biology->term = 'Resting potential';
        $biology->definition = 'Membrane potential';
        $biology->save();

        $biology = new Biologyterm;
        $biology->term = 'Synapses';
        $biology->definition = 'Are Junctions with tiny gaps';
        $biology->save();
  }
```

Glossary View Page:

```
@layout('layouts.default')
@section('content')
<div class="row">
 <div class="span3"></div>
 <div class="span4">
  <img src="/img/neuron.png" Widt="250px" Heght="250px;"/>
 </div>
 <div class="span4"></div>
</div>
<div class="row">
 <div class="span4"></div>
 <div class="span4">
  <h2>Biology Glossary</h2>
 </div>
 <div class="span4"></div>
</div>
<div class="row">
 <div class="span4"></div>
 <div class="span4">
   {{Form::open('biologysearch','POST')}}
   {{Form::token()}}
  <div class="input-append">
</div>
{{Form::close()}}
</div>
<div class="span4"></div>
</div>
@endsection
```

Final Work:

The Academic Glossary Website is accessible through **glossary.pagodabox.com.** The Academic Glossary allows students to have access to their glossary words for each textbook. This helps students to find a word quickly.

Credits:

Cynthia Chan (Biology Teacher In Hoover High School)
Diana Cummins (Dance Teacher In Los Angeles City College)

ACADEMIC GLOSSARY API
AUGUST 2013

Problem:

When I developed the Academic Glossary website, I thought it was fully functional, but when I tried to expand the overall idea, it seemed that the entire website was just a prototype. A prototype was not satisfying for me, so I had to think of something to make the idea more functional and useful.

Idea:

After researching for a while, suddenly I thought of a global service, by which I mean more of a service that all the developers from all over the world can use to develop their ideas and make something interesting and useful with the same database. Basically everyone is going to the same database to add on more glossaries with different languages and develop more sophisticated app, which will be useful in different countries .Foe example, a developer in Australia knows about students in his country and he decides to use the global service and develop an app or website that will be more useful for Australian students. The same concept applies to developers in the Untied States and all other countries.

Inspiration:

When I started to work on Simple-Visa's API, I simply could not hide my happiness. Every single day I was seeing cool feature. Everyday I was using the API to convert our ideas to actions. This is when I thought to do the same thing with my so-called Biology Glossary Prototype. So I borrowed a book by O' Reily from my boss and read it in one day. After that, I got more inspired and amazed.

What is ag-api:

Ag-api is the short name for Academic Glossary API. The main purpose of this API is to provide some functions and actions to handle the Academic Glossary database. This API can be the main scope of each app or website using the same database. It gives the opportunity to different developers to use this API an database in their apps. In some way, all the brilliant minds will help to make amazing apps with amazing ideas with the same database.

Documentation for Non-Developers

Tools for Development of ag-api:

In order to develop the ag-api, I had to use different tools. Some of these tools are used to create and control database in which glossary information will be saved. Some of these tools are used to develop the API functions to perform different tasks and connect the client side to the database through server side.

Initialization for Development of ag-api:

In order to start the development of ag-api, I had to set my workstation so that I could run the codes in specific conditions. Also, I had to set a testing environment to test the codes. It is like solving a math problem but using a different way to check the solution. I had to repeat the concept to make sure all the functions were working correctly.

Connecting to the Database:

I have to say in an API, everything is dependent to the database. All the functions perform tasks on database either to get, update, or delete information. Therefore, it is crucial to check the connection of the database. At the beginning I had to connect to the database with correct credentials. Then I had to test it a few times to make sure the connection was correct.

Constructing and Designing Models:

After establishing a stable connection, it is important to design and construct models, which represent the data plans in the database. Models guide the data to their correct place. Also, models specify what parameters are needed to complete the data sheet or table. For example, in the ag-api, the glossary modes asks for school information, glossary words, and course information. All these fields are specified in the model and every time a user wants to make a new glossary, the inputted data gets injected in the model and saved in the database with the right order and credentials.

Development of API:

A REST API is actually a set of functions about performing tasks on a set of data. These functions are CREATE, READ, PUT, and DELETE. These functions alone build 80% of the API. For instance, in each collection or data set, users can request to read the data. Therefore, the API needs to call the READ method. In order to execute the request, the server side has to process it, so in the API I had to put few lines of code regarding the specified data set or collection with READ, or more specifically the GET method. Usually all of these methods are used in an API. In this case, I used them for both the Glossary collection, and Teachers collections.

Documentation for Developers:

Tools for Development of ag-api:

In order to develop and design this REST API, I had to use certain tools to store the data, control the PAI, and of course develop the API.

Mashape: Mashape is a tool for consuming and distributing API's. I used it to store and consume ag-api. Also, it helps to create test endpoints.

Heroku: Heroku is a hosting provider that I used to host the ag-api.

Node JS: Node Js is a JavaScript server that allows developers to run JavaScript based apps on internet.

MongoDB: MongoDB is an open source document, and noSQL database. I used MongoDB to build a document-oriented database.

Mongoose JS: Mongoose JS in an object modeling tool for Mongo DB. I used it to create schemes for the collections in the database.

Express JS: Express is a web application framework for Node JS. I used this framework because it helps to communicate with the Node server easily, and it is powerful.

Connecting to the Database:

In order to design and develop the ag-api, first of all, I had to connect the database with the help of express JS framework. Fortunately, Heroku gives a lot of add-on apps for the hosted services in their servers. So I added the Mongo Lab to manage the MongoDB databases and use its codes to connect the API to it. In order to do so, I had to run a few lines of code in my main server file.

```
var uristring =
process.env.MONGOLAB_URI ||
process.env.MONGOHQ_URL ||
'mongo-db-url';

mongoose.connect(uristring, function (err,res){
    if(err){
        console.log('ERROR connecting to: ' + uristring + '.' + err);
    }else{
        console.log('Succeeded connected to: ' + uristring);
    }
});
```

Designing and Adjusting URIs:

URI is the most important component in development of REST API. A URI has to be clear, legible, and distinguishable. In order to make the URIs clear, I had to adjust them based on several rules. For example, I had to use hyphens instead of underscores, and lower case letters instead of capital letters. Following these rues helps to establish a clear communication tunnel between the client side and server side.

```
app.get('/glossary/:id', function (req, res) {}
```

In the code above, I simply design a URI for reading a specific glossary in the glossary collection. The "Glossary" in the URI represents the collection and "id" represents the ID of that specific glossary. The Glossary collection itself is a collection of glossary objects. Each glossary that is being created is being saved in this collection.

Another concept of legible URI is in its segments separated by a forward slash. This means each segment itself should represent a resource inside specified models. For example, in the URI above, at first I wrote "Glossary" and then "id". This means that the following URI, which is the first segment of the previous URI, should represent a specific resource. In this case, Glossary segment lists the entire Glossary collection available in the database.

```
app.get('glossary', function (req, res) {}
```

It is important to avoid any use of CRUD function names in URI. CRUD stands for, Create, Read, Update, and Delete. These four actions are the basic and primary actions interacting with the server. These key names should not be used in the URI itself. For instance, in the code below, if the URI is meant for deleting a glossary, the delete function comes before the URI itself.

```
app.delete("/glossary/:id", function(req,res){}
```

Header Request Methods:

In order to write the necessary function for interacting with the server, I had to use the proper header request methods. If these header request methods are misused, there will be complications on the client side while trying to communicate with server side. The four header requests that I had to use in this API are GET, POST, PUT, and DELETE. Each of these headers perform a specific CRUD operation. For instance, if the purpose is to create a new Glossary, the POST method has to be called. If instead of using post I use DELETE, when the user clicks on the save button, the app will try to delete the Glossary and even it might crash.

```
app.get('/glossary/:id', function (req, res) {
GlossaryModel.findOne({ glossaryId: req.params.id}, function (
err, glossary){
      if(err)
      res.send(err);
      else
      res.send(glossary);
   });
});
```

Constructing and Defining Models:

Models are important in the development of an API. Basically without models, we cannot perform any action related to database collections. Models together form collections. However, we have to define the structure of needed models inside the API and create instances of them to perform any actions. Therefore, the first thing that is needed is the scope of the model. The following code shows the model for the Glossary table.

```
var termSchema = new mongoose.Schema({
    word: {type: String, required: true},
    definition: {type: String, required: true}
});

var glossarySchema = new mongoose.Schema(
{
    creator: {type: String, required: true},
    description:{type: String},
    source: {type: String, required: true},
    name: {type: String, required: true},
    course: {type: String, required: true},
    schools: [schoolSchema],
    glossaryId: {type: String, required: true},
    terms: [termSchema]
}
);

var GlossaryModel = mongoose.model('glossary',glossarySchema);
```

CRUD Operations for Each Collection:

CRUD operations provide the main functionality of REST APIs. For each collection I had to use Crud operations to perform actions on the server side. These functions in the API make it easier for client-side developers because they just have to call the functions and provide the proper parameters needed. Most of the time, these parameters are collected from users who input their information to get their desired results.

"C" stands for CREATE operation; this operation uses POST method to create new object in the database. The POST method is used for and operation that needs to send any kind of data to the server. For instance, in order to create glossary teacher objects, I had to use this POST method. However, it is important to provide the proper information to complete the action. If the information that the user is sending Is insufficient or does not match the validation, and error response will be returned. The following code represents CREATE operation with the use of POST method for both Teachers and Glossary collections.

```
app.post("/glossary/new",function (req, res) {
    var glossaryname = req.body.name;
    var courseName = req.body.course;
    var randomnumber = Math.floor((Math.random()*2000)+1);
    var selecedName = glossaryname.charAt(0)+courseName.charAt(0);
    var glossaryid = selecedName + randomnumber;
    var glossary = new GlossaryModel({
        creator: req.body.creator,
        description: req.body.description,
        source: req.body.source,
        name: req.body.name,
        course: req.body.course,
        schools: req.body.schools,
        glossaryId: glossaryid,
        terms: req.body.terms
    }).save(function (err, newGlossary) {
    if (err){
        res.send(err);
        }else{
        res.send(JSON.stringify(newGlossary));
    }
    });
});
```

'R' stands for READ. It uses the GET method to retrieve information from the database. Just like POST, GET operation, based on its purposes, needs some parameters from the user in order to send back response. Sometimes, the GET operation does not need any extra parameters. Usually this is the case when users want to get an entire collection of objects. For instance, to get the entire Glossary collection, I had to use GET method only. However, for individual retrieval, I had to put an extra parameter that is unique for individual glossary object. Usually, developers use the ID parameter generated automatically.

```
app.get('/glossary/:id', function (req, res) {
GlossaryModel.findOne({ glossaryId: req.params.id}, function (err
, glossary){
        if(err)
        res.send(err);
        else
        res.send(glossary);
    });
});
```

'U' stands for UPDATE. Update uses PUT method which is responsible for updating information in an existing model. It is important to provide the unique ID of the model to update the information. If the ID is not provided the PUT function will create a new model.

```
    app.put('/glossary/:id', function (req, res) {
    GlossaryModel.findOne({glossaryId: req.params.id}, function (e
rr, glossary){
            glossary.creator = req.body.creator;
            glossary.description = req.body.description;
            glossary.source = req.body.source;
            glossary.name = req.body.name;
            glossary.course = req.body.course;
            glossary.schools = req.body.schools;
            glossary.glossaryId = req.body.glossaryId;
            glossary.terms = req.body.terms;
            glossary.save(function (err){
                if (!err)
                res.send(glossary);
                else
                res.send(err);
            });
        });
    });
```

In the code above, in order to update a specific Glossary, I used the PUT method at the beginning and the in the URI, I specified the ID parameter '/:' . The ID varies based on which glossary does the user want to see.

'D' stands for DELETE. It uses the delete method itself. In the following code, I simply called the delete method and specified ID parameter, so it will delete only the desired object.

```
app.delete("/glossary/:id", function(req,res){
GlossaryModel.findOne({glossaryId: req.params.id}, function (err, glossary) {
        glossary.remove(function (err) {
            if (!err) {
                return res.send('');
                } else {
                console.log(err);
            }
        });
    });
});
```

Consistent Error Responses:

Response errors are important for users. If users are getting errors, they expect them to be clear and informative. Therefore, it is important to use consistent and clear errors based on their type.

Response Body:

Response body is the response that is shown to the user based on different requests. There are different response bodies for different requests. Therefore, I had to make sure that they are well developed and organized. They have to be clear and concise. In APIs, response bodies should support JSON. JSON helps developers to access data as objects. Therefore, it is important to have JSON as a response body.

```
{
    "creator": "arpiderm@gmail.com",
    "description": "Biology Glossary For Sheltered Section",
    "source": "Preason Hall Publications",
    "name": "Sheltered Biology",
    "course": "Biology",
    "glossaryId": "SB246",
    "_id": {
        "$oid": "522bbb3026c4bc0000000009"
    },
    "terms": [
    {
        "word": "Test",
        "definition": "Definition2",
        "_id": {
            "$oid": "522bbb3026c4bc000000000c"
        }
    },
    {
        "word": "Test2",
        "definition": "Definition2",
        "_id": {
            "$oid": "522bbb3026c4bc000000000b"
        }
    }
    ],
    "schools": [
    {
        "name": "Hoover",
        "state": "CA",
        "city": "Glendale",
        "zipcode": "91205",
        "_id": {
            "$oid": "522bbb3026c4bc000000000d"
        }
    }
    ]
}
```

Not only should JSON be supported in response bodies, but has to be extremely organize and legible. For example, in the glossary model, all the parameters for course information should be in one place on after another.

Final Thoughts:

Having an API for Academic Glossary is a great accomplishment. It opens a lot of doors to develop powerful apps for students and teachers. I want to thank to everyone who inspired and gave me the power to develop this REST API. With the development of this API, I am taking one step further and getting closer to a more sophisticated development world.

ACADEMIC GLOSSARY APP
SEP 2013

Problem:

The problem that I was facing started when I finished the Academic Glossary website. When I started to use the website to check its functionality, I realized that websites are not accessible every where. People need to access the Internet and have a relatively big screen to be able to use the Glossary while away from home or a computer. The main point behind this idea was to have a fully accessible App for people, so they can have access to it in the easiest way and also suggest new glossary sources. Unfortunately with the website, it was not possible. Therefore, I had to look for a better solution for this problem. As a matter of fact, I needed a better Idea to solve a problem within a problem. I guess this is one of the problems that I would call it nested problems just Like "Nested If statements" in programming.

Idea:

All those observations were on hold until I started to work on the SimpleVisa ISO APP for MINEA USA Company. After developing and releasing the first version, I realized that people started to use the app and get information about their VISAS even more than our website services. I realized that our app helped a lot of people in airports to either renew their Electronic Pass or to make sure of their expiration date. They were using the app because before, they would be stuck in the airport because their ESTA had expired and they could not fly to the USA without that ESTA. Seeing that much of traffic on our app from the first deploy, a clear Idea came to me, to do the same thing for the Academic Glossary. This app will allow people to look for technical words while in a rush in the easiest way. From my heart I knew that this would change the way people threated glossary words in the back of the books that they read.

"What If" and its Power to Solve Problems

Inspiration:

The inspiration for converting this idea to action came from the challenge that I created myself. I agreed to develop an app without knowing a single thing about developing IOS or mobile apps. But I did not give up. I started to stay up at night and learn from observing other source codes. In the morning I would go in the office and try everything myself. After a while, I noticed that just with trying and learning I have created the basic scope of the SimpleVisa app. That scope was a start up for a new change for the development branch of our company. After that challenge I started to extend our app and implement advanced codes to make it even more powerful and in the meantime much simpler. This was the inspiration. The result of this app was the inspiration and source of a huge energy for me to work on the Academic Glossary App, also known as (AG-APP)

Blueprints of the AG-APP:

A basic step-by-step plan for a successful project is crucial. Therefore, after sketching up the app on paper, I developed a basic plan according to the sketches. In that way, at least there will be a guide, which will help to decide from where to start and where to end. The following plan represents the flow of this app.

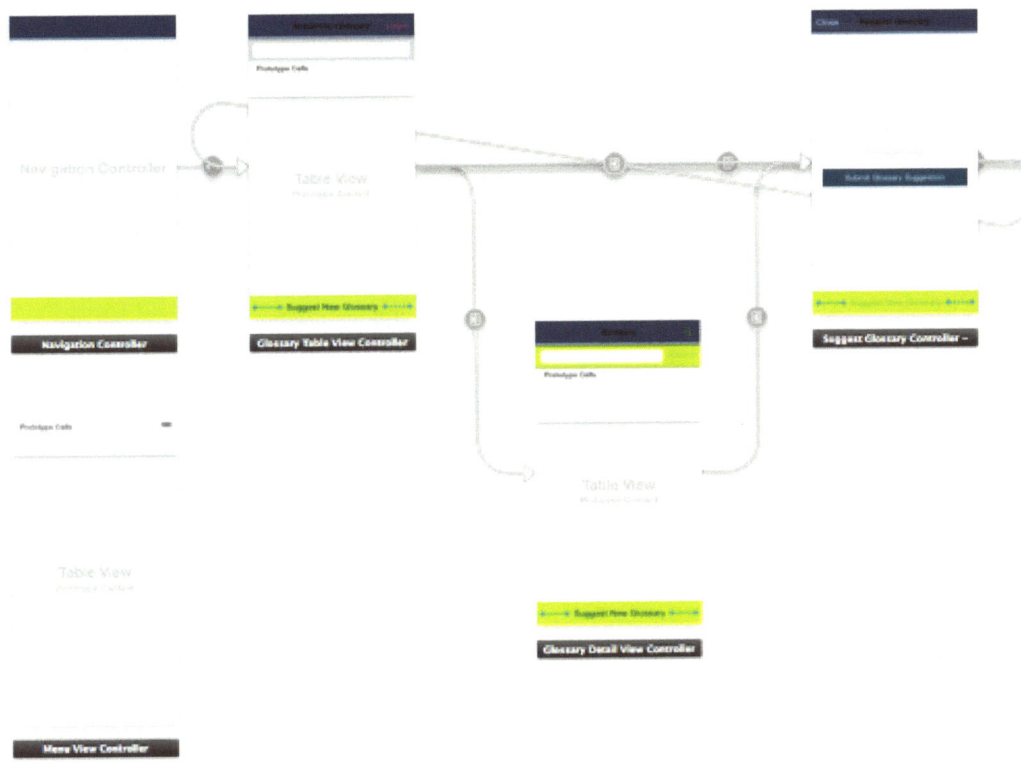

Documentation For Non Developers:

After an idea comes to us, it is very crucial that developers design a mockup or wireframe or even a simple sketch up of the app on paper. Paper and pencil is the starting point of every single project, app, or website. In all the experiences that I have had, I have learned that paper and pencil work magic. I have learned, that all the details come easily on paper. And I learned that a good construction of any app has to have a strong base, which starts from all the sketches on the paper. Therefore, drew the app many times. Each of the sketches had some details that the others did not have. So I had to combine all those sketches and create a master sketch, which would become the initial model for the development of Academic Glossary App.

Resources:

Resources compose an important part of any project. Without resources it will be extremely hard to finish a project; especially if a project relies on other sources. For example, for The Academic Glossary I had to ask permission from a lot of professors, publishers, or writers in order to list their Glossaries in the AG-APP. Also, for the development itself, I needed graphic files. Not many developers design the user interface by themselves. So I had to either buy some graphic pieces for the app or ask someone to create them.

Technology Used:

Before the development of an app, it is very important that developers decide what technology is going to be used. By technology I mean the programming language, development environment, and all the resources belonging to that language. For instance, for this app I decided to use IOS SDK, which is Apple's unique development environment. IOS SDK uses Objective C programming language, which again belongs to Apple.

Tools:

There are many developer tools out there, but not all of them can be used for desired purposes. For IOS App development, I had to Use XCODE app, which can be found in Apple Store.

Developing The App Based On Sketches in Order:

In order to develop this app without missing any detail, it is important to follow the plans and blueprints.

Create List Of Courses:

Based on the project plan and the sketches, creating list of courses is the first and most important point that will give the app a new dimension and a new face. In this step basically I had to develop a list that contains all the courses in the database. This list shows Glossary Name, Glossary Unique ID, and name of the school. This information is being generated in separate list items. As I said earlier, the main point of this app is to be simple and powerful. Therefore, it is important to list available Glossary for all the courses. In this way, users will be able to choose the desired glossary.

Link to each glossary Terms List:

If the user chooses a course, he needs to get access to the Glossary Terms which belong to that course. Therefore, it is crucial to create a link on each Course Item that leads to the Glossary Terms page belonging to that specified Course. The link or "Segue" is like a link on websites that leads to another page, or lead to a detail page for a specified word or term or phrase. The concept is the same. It only changes based on different purposes and needs.

Search Controller To search Through Course:

It is true that listing all the courses at the beginning will help users a lot. However, there is a complication that will occur in different cases based on users' desires. Therefore the app needs to be prepared for any case and any complication. One of these cases is that the user will need to look for a glossary name that starts with S, for example. At this time, without the possibility of searching, the user will have to scroll down to get to the S section. This will be not simple for users anymore. Therefore, it is important to prepare the app to react based on users' desire and intentions.

Individual Glossary detail View Page:

List Of Glossary Terms belonging to the selected glossary:

After selecting the desired glossary for a specific course, the user should see the glossary terms immediately. Therefore, when the page loads, the most important task of the app will be to get the related glossary terms from the database and show the to the user in a list table view.

"What If" and its Power to Solve Problems

Link back to list of courses:

Navigation is the most important aspect of phone apps. Users should be able to use the app without getting lost. Therefore, just like back button on web browsers, apps should provide back buttons on navigation controller so users will feel safe. Also, it is important to label the back button with the title of the previous page that the user came from.

Search controller to search through glossary terms:

Just like searching through courses, it is important that users have the ability to search through the glossary terms. The whole point of this app is to help people find the desired glossary terms without any hassle, without having to turn the book pages over and over again to find a single word.

Suggest Glossary Page:

It is obvious that from the beginning this app will not have all the desired courses. Therefore, I have had to implement the "suggest glossary" feature, so the users will be able to suggest glossaries related to their courses. Only at that time, can our app implement those courses with their Glossary terms.

Finalizing The App:

When the main development of the app is finished, it is crucial that we test the app and look for any issues. Also, it is important to choose a single color scheme and stick with it through the app and its release process. Surprisingly, the choice of colors and art of designing the app motivates users to try the app with joy and complete satisfaction, knowing that in one aspect their life just got easier.

Documentation For Developers:

Technology Used:

For the development of this app, I used Objective-C programming language and X Code apple developer environment. Before using Objective-C, I was considering using Phonegap with Java Script. However, later on I found out that for Apple devices it is better to use Objective C so I could have a complete control over targeted devices.

Tools:

The major tools that I have used for this app are, Adobe Illustrator for the wireframes, icons, and graphics; XCode for writing codes; and Parse for handling the back end service.

Database:

For this app unlike my other projects, I decided use something new for my database. Therefore, I decided to use no SQL, or document oriented, or non-relational database. Fortunately the PASE backend service has already integrated this kind of database in it. Therefore, I decided to store all my data in PARSE. The most important advantage about PARSE is that is supports multiple platforms. Therefore, if I decide to develop the same app in different platforms, I will not have any problems with reusing my database.

Developing The App Based On Sketches in Order:

Create List Of Courses:

The first and most important job of the app is to retrieve all the courses from the database and list them in alphabetical order in a table view. When users first enter the app ,they should see this list immediately. The following code block shows the function that does this job in the home page

```objectivec
- (void) glossaryTable {
    
    PFQuery *retrieveGlossary = [PFQuery queryWithClassName:@"Glossary"];
    [retrieveGlossary findObjectsInBackgroundWithBlock:^(NSArray *objects, NSError *error) {
        if(!error){
            glossaryArray = [[NSArray alloc] initWithArray:objects];
        }
        
        [glossaryListTable reloadData];
    }];
```

Link to each glossary Terms List:

When the user selects a desired Glossary, the app should pass the specified glossary unique id to the glossary terms view page. We should pass in the ID so the app will know which glossary terms to fetch from the database. In objective-c there is this method to pass in data from list item to another page. This method uses segue linking feature. The code below demonstrates how the code is passes info using segue linking.

```
(void) tableView:(UITableView *)tableView didSelectRowAtIndexP
ath:(NSIndexPath *)indexPath{
    [self performSegueWithIdentifier:@"showGlossaryTerms" send
er:indexPath];
}

(void)prepareForSegue:(UIStoryboardSegue *)segue sender:(id)se
nder {
    if([segue.identifier isEqualToString:@"showGlossaryTerms"]
) {
        NSIndexPath *indexPath = [self.tableView indexPathForS
electedRow];
        GlossaryDetailViewController *glossaryDetailViewContro
ller = segue.destinationViewController;
        PFObject *details = [glossaryArray objectAtIndex:index
Path.row];
        NSString *objectId = details.objectId;
        NSString *name = [details objectForKey:@"name"];
        glossaryDetailViewController.glossaryID = objectId;
        glossaryDetailViewController.glossaryName = name;
    }
}
```

Search Controller To search Through Course:

In order for users to be able to search through the courses, the app should handle the search though importing the filters into a new array and render the table view. This is done in multiple steps and methods. The codes below shoe these steps.

```objc
self.searchBar = [[UISearchBar alloc] initWithFrame:CGRectMake
(0,0, self.view.frame.size.width, 44)];
self.tableView.tableHeaderView = self.searchBar;

self.searchController = [[UISearchDisplayController alloc]init
WithSearchBar:self.searchBar contentsController:self];
self.searchController.searchResultsDataSource = self;
self.searchController.searchResultsDelegate = self;
self.searchController.delegate = self;

CGPoint offset = CGPointMake (0, self.searchBar.frame.size.hei
ght);
self.tableView.contentOffset = offset;
self.searchResults = [NSMutableArray array];

- (void) filterResults: (NSString *) searchTerm {
    [self.searchResults removeAllObjects];
    
    PFQuery *searchQuery = [PFQuery queryWithClassName:@"Gloss
ary"];
    
    [searchQuery whereKeyExists:@"name"];
    [searchQuery whereKeyExists:@"glossaryID"];
    [searchQuery whereKey:@"name" containsString:searchTerm];
    
    
    NSArray *results = [searchQuery findObjects];
    [self.searchResults addObjectsFromArray:results];
    
}
```

Individual Glossary detail View Page

List Of Glossary Terms belonging to the selected glossary:

After selecting a specified Glossary and passing the unique identifier with the use of segue, the app should fetch all the glossary terms belonging to that Glossary and show it to the user in alphabetical order.

```objectivec
- (void)glossaryTermsTableFetch{
    
    PFQuery *retreiveGlossaryTerms = [PFQuery queryWithClassName:@"Glossary"];
    [retreiveGlossaryTerms whereKey:@"objectId" equalTo:glossaryID];
    
    [retreiveGlossaryTerms getFirstObjectInBackgroundWithBlock:^(PFObject *glossary, NSError *error) {
        
        if(glossary){
            
            NSString *objectId = glossary.objectId;
            PFQuery *query = [PFQuery queryWithClassName:@"glossaryTerm"];
            [query whereKey:@"glossaryID" equalTo:objectId];
            [query orderByAscending:@"term"];
            [query findObjectsInBackgroundWithBlock:^(NSArray *objects, NSError *error)  {
                if (!error) {
                    glossaryTermsArray = [[NSArray alloc] initWithArray:objects];
                } else {
                    //========== Log details of the failure
                    NSLog(@"Error: %@ %@", error, [error userInfo]);
                }
                [glossaryTermsTable reloadData];
                
            }];
        }
        
        [glossaryTermsTable reloadData];
        
    }];
}
```

Search controller to search through glossary terms:

Just like searching through the courses, the app should handle the search of glossary terms in an efficient way. This process is the same as the one mentioned earlier. However, it is important to notice that this time, it will use a different class name and different array names.

Suggest Glossary Page

Suggest Glossary Page handles the submission of suggesting new glossary form. It generates a form that users have to complete with information regarding the desired glossary. After submitting the form, the app saves the information in a separate class in Parse service, so our developers will be able to add new suggested glossary for different courses in the app.

Finalizing The App

Before releasing the app, it is very important to debug and test it. It is required to get approval of other developers in the group to omit any syntax errors. Usually for testing the app, the best option is to use Test Flight. With test flight test users can try the app and look for any problems or suggest improvements. This pre release step solves many problems. Also, with Test Flight users will be able to actually test on their devices, which is considered an advantage.

"What If" and its Power to Solve Problems

Final Work

The Academic Glossary IOS app can be found in the Apple App Store or on our website called www.academicglossary.com. The following screenshots show the preview of the app on IOS devices.

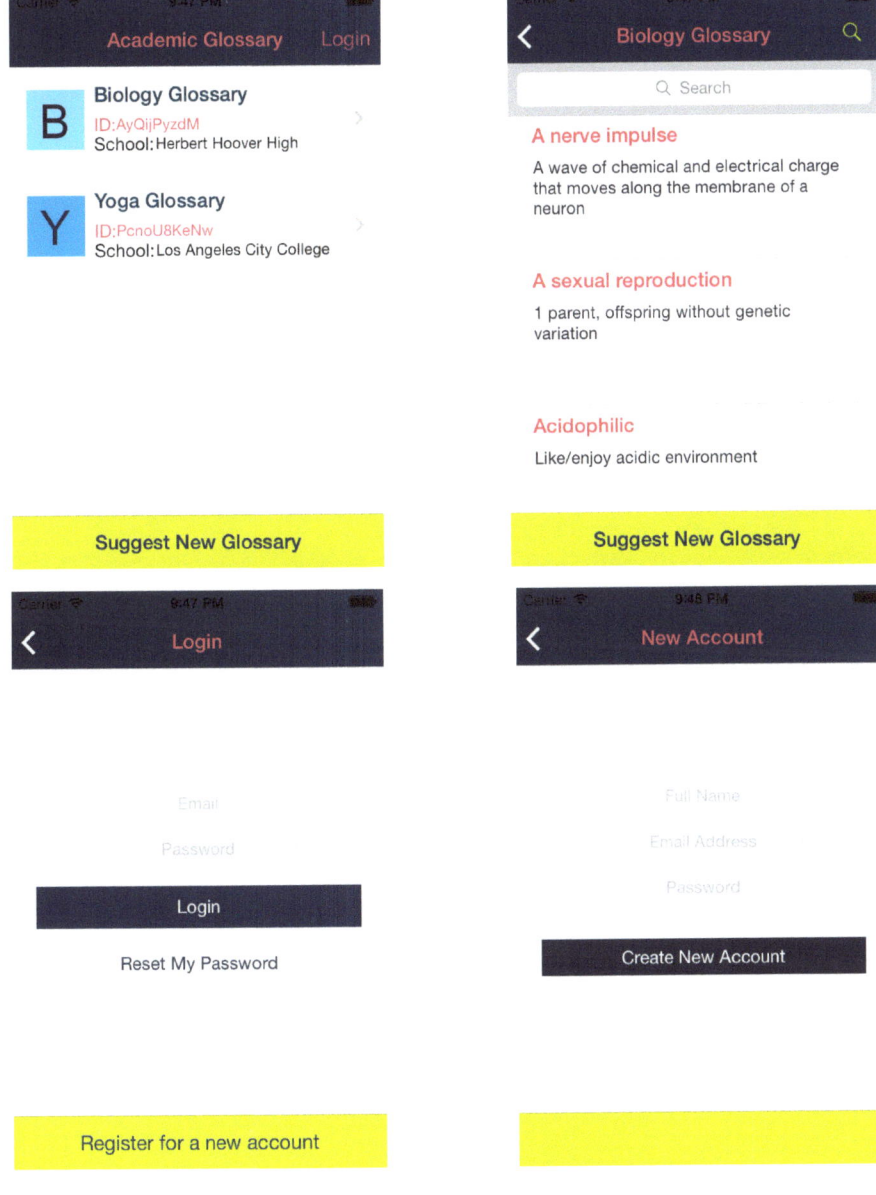

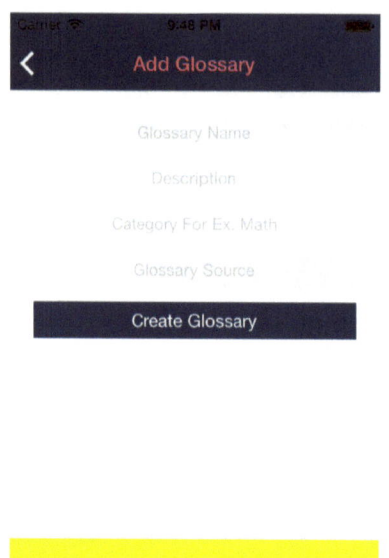 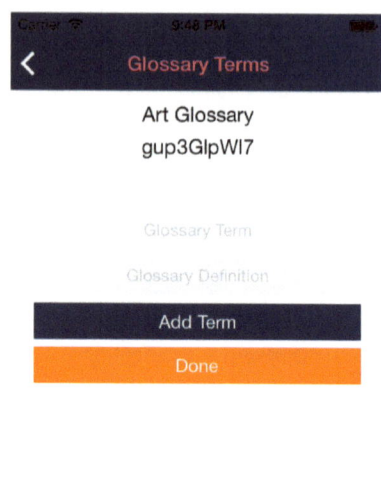

Further Goals

Usually, after converting ideas to actions, we keep improving our ideas. We start to imagine our ideas in more complex and ideal forms. Therefore, we keep changing and improving them no matter what. In app development, this imagination of improving our ideas and helping them to be come ideal is amazingly helpful and effective. This improvement of ideas helps to improve the app itself and finally to make users even more satisfied. Therefore, with the use of my imagination and desire to reach to ideal form, I developed a list of further features and improvements of the app. This list includes the following features.

1. Teachers' Login System
2. Adding Glossary Directly from the App when logged in.
3. Sharing desired courses
4. Listing courses based on school, cities, or states.
5. Addling language options for users in different countries.

Credits

In order for me develop this few people helped. I have to thank my boss at Minea USA, Loris Mazloum for the inspiration and professional advice. For the glossary terms, I have to thank Cynthia Chan, Diana Cummins, and Json Pinsker. Also, I the Biology Glossary terms are taken from Pearson Prentice Hall's Biology book.

Final Words

And here we are again, the typical conclusion at the end of each project. Except that this one will not be typical anymore because I decided to fill it with the words that come from my heart. How difficult is it to finish an entirely technical chapter with the words from my heart. Well, here is the truth. Everything starts from the centre and everything ends at the centre of my heart. These ideas come from my observations of the problems that we face on a daily basis. These ideas rise and inspire because in my heart I think of helping to solve this problem with the use of technology, and science. Without being able to help, my knowledge will not be useful. When the idea and inspiration rises, my technical side turns on and busts up, so I can convert the idea to a solution. Until I solve the problem, my technical side stays on. At the end, my heart starts to feel satisfaction, joy, and inspiration. Yet, it never feels satisfied, because before even I taste that satisfaction my brain starts to take snapshots from other problems that we are facing on daily basis. This is it. This is why we live on this earth. This is why we must never stop. This what makes us happy, and this is what keeps us alive: learning, experiencing, finding problems, and solving problems.

ABOUT THE AUTHOR

Arpi Dermardirousian was born in Isfahan, Iran. Until age of 13 she was fascinated by anything related to electronic components. At age 13, she started to fix computers. By the time she was 15, she developed her first windows application regarding to a problem that she was observing on daily bases with the help of an experienced developer. Soon, she sent her application to the Nation Wide Kharazmi Competition in Iran. This app which was called "Speak True", became her first innovation. With this app she was able to suggest a logical solution to one of the long lasting problems surrounding Armenians in Iran. Perhaps, this application guided Arpi throughout her life and made her who she is today. After she moved to USA, she never stopped looking for problems surrounding her, and adopting her abilities to solve those problems. Currently, she is majoring in Software Engineering field of study, and works for an electrical manufacturing company as their developer and programmer.

www.ingramcontent.com/pod-product-compliance
Lightning Source LLC
Chambersburg PA
CBHW050859180526
45159CB00007B/2724